ACIDS and BASES

Rebecca Woodbury, Ph.D., M.Ed.

Gravitas Publications Inc.

Acids and
BASES

Illustrations: Janet Moneymaker

Copyright © 2025 by Rebecca Woodbury, Ph.D., M.Ed.

Acids and Bases
ISBN 978-1-950415-12-0

Published by Gravitas Publications Inc.
Imprint: Real Science-4-Kids
www.gravitaspublications.com
www.realscience4kids.com

RS4K

Image credits: Cover & Title Pg, P. 15: Soda, By ozmen, AdobeStock; Cover & Title Pg, P. 16: Banana, By xamtiw, AdobeStock; Cover & Title Pg, P. 16: Cleaner, By Crystal de Passillé-Chabot on Unsplash; Cover & Title Pg, P. 15: Grapes, By Merethe Liljedahl from Pixabay; Cover & Title Pg, Above & P. 6, 13, & 15: Lemons, By Varintorn Kantawong from Pixabay; Cover & Title Pg, P. 16: Dates, By olyina, AdobeStock; Cover & Title Pg, Above & P. 6, 13, & 16: Soap, Alecsander Alves on Unsplash; P. 3. By Seika, AdobeStock; P. 15, Car Battery, By warloka79, AdobeStock; Dry Cells, By New Africa, AdobeStock

What happens when you eat a lemon?

Do your lips pucker?
Does your tongue tingle?

Lemons taste **sour!**

Pucker up!

Not me!

What happens when you grab
a bar of soap?

Does it jump out of your hands?

Does it wiggle away when you
try to catch it?

Soap is **slippery**!

Cheese is
not slippery.

Why are lemons sour?

Why is soap slippery?

Lemons and soap are different.
Lemons and soap are made of
different kinds of **molecules**.

I want lemonade.

Review: ATOMS

- **Atoms** are tiny building blocks that can link together.

- **Atoms** make everything we see, touch, taste, and smell.

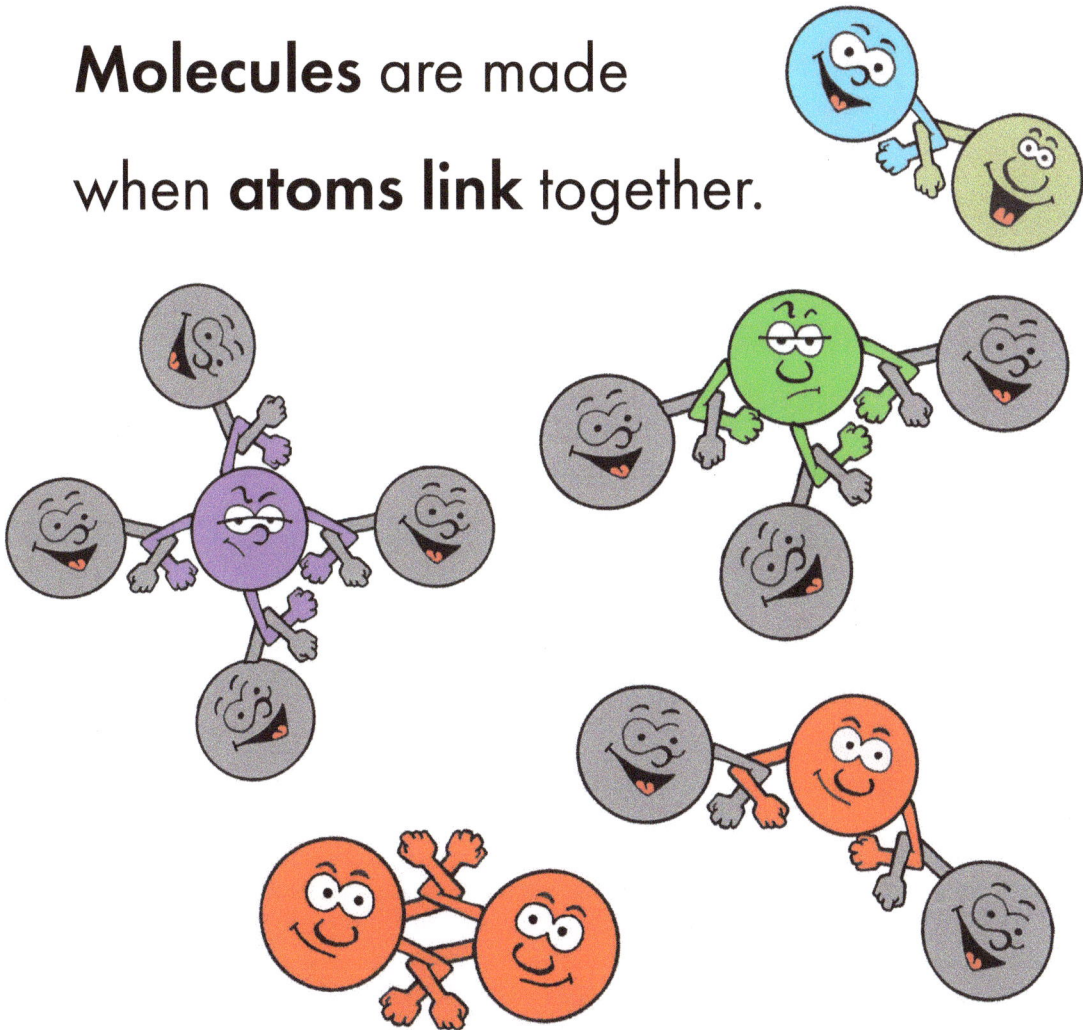

Review: MOLECULES

Molecules are made

when **atoms link** together.

Review: MOLECULES

Molecules can be big or small.

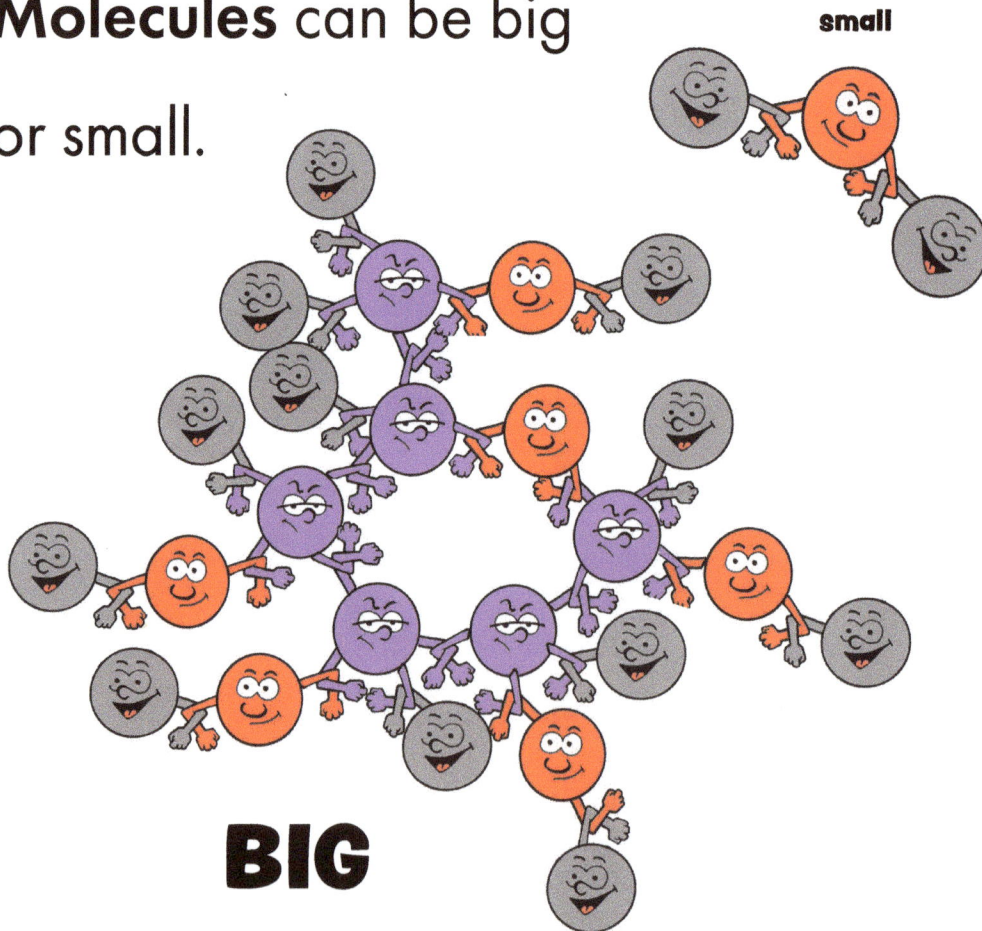

small

BIG

Molecules have different **properties**.

Properties describe what something is like.

Is fur a property?

Slippery and sour
are properties.

Hot and cold
are properties.

Hard and soft
are properties.

- Lemons have molecules that are sour.

- Soap has molecules that are slippery.

- Sour molecules are often **acids**.

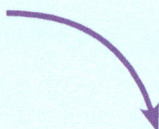

Acids and bases are two different types of molecules.

- Slippery molecules are often **bases**.

Do you use soap?

Never!

Acids are found in batteries, lemons, grapes, and soda pop!

Bases are found
in cleaners, soap,
bananas, and dates!

Acids have an **"H"** group.

An "H" group is a **hydrogen atom**.

Hi! You can find me in acid molecules.

Hydrogen atom

An **"H"** group is part of an **acid molecule.**

H group

We are an acid molecule.

Hydrogen atom

Chlorine atom

Molecule

Bases have an **"OH"** group.

An **"OH"** group is an **oxygen**

atom and a **hydrogen atom**

hooked together.

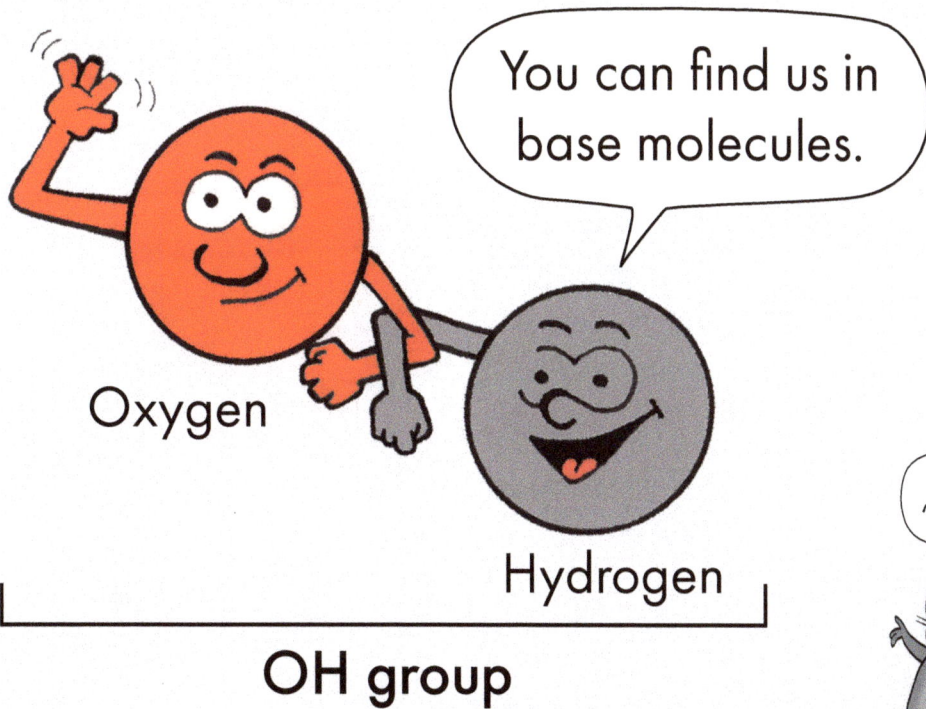

Oxygen

You can find us in base molecules.

Hydrogen

OH group

Say "O" "H"

An "**OH**" group is part of a **base molecule.**

We are a molecule that is a base.

OH group

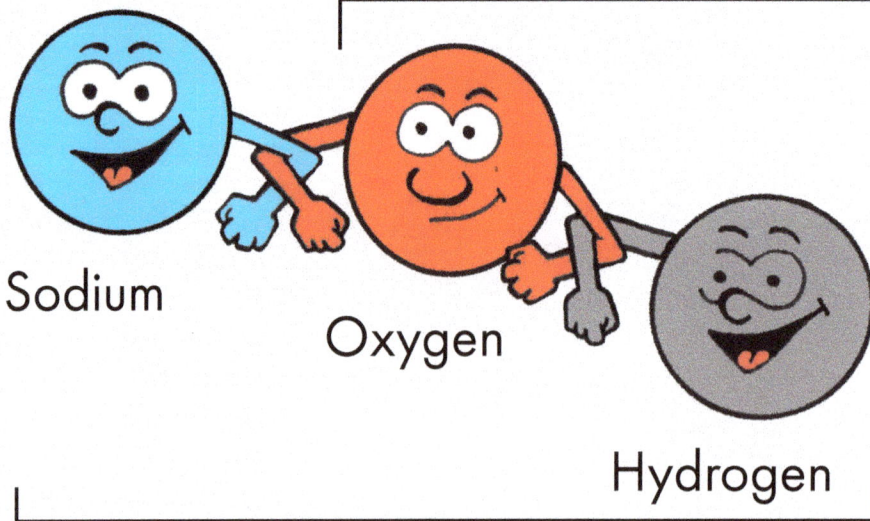

Sodium

Oxygen

Hydrogen

Molecule

Acids and bases are different. Acids and bases have different atoms in their molecules.

Acids have an **"H" group.**

Bases have an **"OH" group.**

"H" groups and **"OH" groups** give acids and bases different **properties**.

How to say science words

acid   (AA-sed)

atom   (AA-tum)

base   (BAYSS)

chlorine   (KLAW-reen)

H group   (AYCH groop)

hydrogen   (HIY-druh-juhn)

molecule   (MAH-lih-kyool)

OH group   (pronounced: "O" "H" groop)

oxygen   (AHK-sih-juhn)

property   (PRAH-puhr-tee)

sodium   (SOH-dee-um)

www.ingramcontent.com/pod-product-compliance
Lightning Source LLC
Chambersburg PA
CBHW040148200326
41520CB00028B/7533